# Agate Hunting Made Easy

By Jim Magnuson

Photographs by Carol Wood

**Adventure Publications**
**Cambridge, Minnesota**

# Dedication

In loving memory of our father Norman Magnuson whose deep love and respect for nature, and passion for educating his children about his native north woods, inspired us all to make a connection with the wonders of God's creations.

Also, a special thanks to our sister Ann O'Keefe for her assistance and support in developing this book.

Many thanks to Ron Whealdon for providing first-hand instruction on agate hunting in rivers and streams. With his guidance, we found several agates and know that you'll benefit from his experience and tips.

Book design by Jonathan Norberg and Lora Westberg

15 14 13 12 11 10 9 8

**Agate Hunting Made Easy**

Published by Adventure Publications
An imprint of AdventureKEEN
310 Garfield Street South
Cambridge, Minnesota 55008
(800) 678-7006
www.adventurepublications.net

Printed in U.S.A.
ISBN 978-1-59193-326-7 (pbk.)

# Agate Hunting Made Easy

# Table of Contents

# Introduction

**MANY PEOPLE WHO ENJOY THE OUTDOORS** also enjoy hunting for beautiful flowers, stones, birds and other natural objects. Agate collecting is a popular pastime, and dozens of different agate varieties exist worldwide. The Lake Superior agate is one of the most prized and celebrated types of agate. A banded form of chalcedony, Lake Superior agates get their name because they formed in the lava masses that make up the Lake Superior Basin.

People have been collecting Lake Superior agates for hundreds of years, and likely even thousands, as evidence has been found that Native Americans used high-quality specimens to make jewelry and ceremonial objects. On top of being beautiful, Lake Superior agates have significant financial value; some high-quality gemstones are worth hundreds and even thousands of dollars. Yes, Lakers are awesomely beautiful and valuable, but they are also **rare and elusive**, and many people give up hunting out of frustration. Therefore, it's our goal to make it simple, fast and fun for you to search for agates.

The information in this book is the result of years of learning things the hard way (i.e., by experience), and reading numerous books regarding

Lake Superior agates and their geology. We are going to teach you what to look for out in the wild because the beautiful pictures you see in books and on the Internet are not representative of what rough stones look like in the field. Consider this your **just-in-time reference guide** that you can crack open the week before you head out on your family vacation, or the night before you hit the hunting trail if you are already in a locale that has Lakers.

We break the process down into several easy sections and use up-close, high-quality photographs that leave nothing to the imagination. These visuals will reinforce the written information and put you on the fast track from newbie to successful Agate Hunter.

Our goal is to help you come home with a bucket full of beautiful stones. We'll also show what to do with your finds once you've got them in hand, including how to polish and tumble them and set up appealing agate displays that your friends and neighbors will enjoy and envy!

To help you spot Lake Superior agates, each agate trait in this section includes a pair of images. In one, we show the back side of the stone, which is what you're likely to see. In the next, we show the banded side of the stone.

# The Real Deal: Identifying Lake Superior Agates

## Lake Superior Agates in the Rough

When agate hunting, it's important to do your homework. So pay close attention to this next section—and especially to one piece of advice: As you begin to find actual specimens that exhibit the agate characteristics that we mention, keep the agates in their rough state (do not put oil on them or polish them). Then, put these agates into a Ziploc bag and keep it handy. Get the stones out the night before you go hunting; look them over carefully at the breakfast table in the morning. Be sure to look at the **back sides** of these stones, as this is what you're likely to see when you're hunting. While probability would suggest that you'd be equally likely to find an agate face up, at many hunting locations, stones are dusty/dirty, and it has been our overall experience that 60–70% of agates we find are face down! More importantly, an agate's back side often looks appreciably different from the "banded" side. To put it simply: If you focus only on banding, you won't have as much success.

Even when one finds a Laker face up, it's often easier to identify an agate based on features like pitting, staining, semicircular fractures and limonite staining. So before you leave home, review this section, and when you're hunting, stay on the lookout for these characteristics. This is your key to finding agates in what are often dry, dusty conditions. Your subconscious agate-hunting brain will keep these images in your short-term memory and pull them up just in time! This is one of the best pieces of advice we have to offer, so do your homework on this material again and again.

*Before you go hunting, examine some rough agates so you know what traits to look for.*

back side
banded side

## Common Agate Feature #1

### Round Edges

- Unless damaged or broken, 95% of Lake Superior agates have rounded edges.

back side
banded side

## Common Agate Feature #2

### Pitting

- Most agates exhibit pitting on their exterior surfaces and it's unusual to find a Lake Superior agate without pitting.
- Pits vary from pinhead size up to ¼". Larger pits are normally nicely rounded or spherical, but smaller pits sometimes resemble dimples or make the skin look rumpled. Some pits are squared off or occur at sharp angles. These occur in agates that formed within air pockets in rock. These pockets often already contained crystal formations. The agate formed around these existing crystals, leaving angular impressions in the outer surface of the agate.
- When an agate's exterior has a rough texture and dimples/ rumples, the outer skin is often said to resemble the skin of an old potato.

back side
banded side

## Common Agate Feature #3

### Staining

- An agate's exterior surface is rarely pristine. Aside from pitting, agates are often stained a mustard yellow. Such stains are caused by limonite, an iron-bearing mineral. Other agates are stained in hues of red because of other iron minerals.

- When you see a stone that exhibits pitting and is stained, your heart will race!

back side
banded side

## Common Agate Feature #4
### "Waxy," Translucent Skin

- When you're hunting in the best of conditions—bright midday sun—light will pass through the hard surface of a Lake Superior agate, creating a "waxy" glow. If the skin of a stone has a waxy glow, that's an outstanding clue that it may be an agate, but it may also be a signal of other very hard stones, such as chert or quartzite. Don't hesitate to turn these stones over, especially if the stone shares the Lake Superior agate's characteristic coloration, such as deep red or gray.

- Translucence may only be evident on sunny days, but almost every good-quality Laker has a degree of translucence. If you are hunting on a cloudy day and the sun comes out later, consider revisiting some of the places you previously hunted; you'll be surprised at what turns up!

back side
banded side

## Common Agate Feature #5

### Fractures that Resemble "Half Moons"

- Minerals that consist of quartz often fracture in a semicircular/half-moon shape. Stones with a lower hardness typically break at a flatter or sharper angle. Such fractures are technically referred to as "conchoidal." (The term conchoidal gets its name from the conch shell.) In some cases, you will see small circular whorls and fractures that may exhibit a "ripple effect" of sorts.
- It is also common to see a waxy glow within these half-moon fractures.

back side
banded side

## Other Agate Characteristics

### Banding and Patterns

- While all agates exhibit banding, it's often difficult to see banding when you're hunting amid dusty, dirty stones. (That's why we think it's helpful to look for the other characteristics first, as they are easier to spot.)

- Once you find (and clean) your stones, banding becomes more obvious. The most common type of agate banding is referred to as fortification banding. Most agates you find will be fortification agates. Fortification agates get their name because they resemble forts or walls that are viewed from above.

- Agates with fortification banding have concentric rings that are usually present in a repeating pattern of alternating colors, but bands can also be clear. Anytime you see a repeating pattern on the face of a stone, it's an agate suspect and you should pick it up for further inspection.

back side
banded side

## Color Variations

- Lake Superior agates come in an endless variety of color combinations, but look for reds and rust-browns first, as Lakers are often found in these colors. After the reds, gray or rootbeer-brownish colors are also quite common.
- Rarer colors also exist. One of the most treasured and rare types of agate is called a paint agate. Paint agates feature striking deep orange banding that alternates with colored bands of dark brown, blue, green, yellow or deep red.
- Some Lakers also have a "rainbow outer husk"—you will see reds, grays, purples and other colors that grab your attention, especially on bright sunny days.

back side
banded side

## Peeling

- As agates formed a long time ago, they have been extensively weathered and have been subject to eons of wind and waves, not to mention the natural freeze-thaw cycle each year. These natural forces often cause agate layers to break or erode away, giving an agate a "peeled" appearance. In fact, in such cases, that's really what has happened, and each "peel" exposes the layer beneath it.

back side
banded side

## Quartz/Crystal

- The interior of a Lake Superior agate is often filled with quartz. Quartz or crystal within a Lake Superior agate is generally an undesirable feature, but it depends on how much crystal fill there is and whether there are alternating or "floating" fortification bands interspersed with it, and whether those bands have striking colors.
- Because quartz material is translucent, and because it is generally chalky white against dark red agate bands, it will catch your eye as a visible feature of Lake Superior agates.
- Quartz crystal that is within Lake Superior agates can generally be differentiated from bright white quartzite, which is less translucent.
- Some stones have smoky quartz at the surface.

# Agate Sizes

Hunting for Lake Superior agates is just like fishing—it's a lot of fun to go after the lunkers, but panfish make a great meal and are also enjoyable to catch. You'll be amazed by the beauty and quality that you'll find in even the very smallest of Lake Superior agates!

Unless you have access to a large volume of medium-to-large whole stones, and you are having outstanding success finding larger stones, you should reserve the last hour of the day for hunting "smallies." By the end of the day, your feet will thank you and you will find it peaceful to sit on the beach or on a gravel pile of small stones and hunt for little gems. You'll find out later that even small agates with little or no visible pattern serve a valuable purpose, as you can use them to tumble larger agates.

| Weight | Approximate Size | Proportion of Total Agate Distribution |
|---|---|---|
| 1 lb. and larger | Softball | 1 in 10,000 |
| 8–14 oz. | Baseball | 1 in 1,000 |
| 5–7 oz. | Golf ball | 1 in 500 |
| 2–4 oz. | Ping pong ball | 1 in 20 |
| Smaller | Marble | 95 of 100 |

Agates of various sizes

# Lake Superior Agate Types

By now you have a good idea of what a Lake Superior Agate is, but unless you know about all the different types of Lakers, you are going to throw away some precious and rare agates. This section will teach you about the twelve most common types of Lake Superior agates (yes, there are even more) and provide you with a simple guide to classifying them. By the way, it's common to find an agate that exhibits more than one of these "agate type classifications," so label it as whichever one seems most prominent to you.

## Fortification Agate

- The most common agate variety, fortification agates have multiple concentric (circular) bands with varying colors. Generally, there are two alternating colors, such as red and white, but several colors can be interspersed as well. Agates with a variety of colors, wide bands and striking "color separations" are more beautiful and valuable. While there are many colors in the Lake Superior agate "rainbow," when alternating white or clear bands are present, you are likely looking at a fortification agate.

## Tube Agate

- Tube agates have stalactite-like tubes that either appear as linear features or spiky patterns. Sometimes, the tubes in these agates erode more slowly than the rest of the agate, creating bubble-like features on top of an agate.

Fortification Agate

Tube Agate

Eye Agate

Moss Agate

## Eye Agate

- Agates with one or more perfectly circular bands on the surface. These eyes are actually hemispheres that go deeper into the host agate material. The larger and more pronounced the eye, the more valuable the agate. Some eye agates also have concentric eyes (eyes inside of eyes).

## Moss Agate

- Moss agates have highly irregular banding patterns that are more lacy and interwoven than fortification banding, giving them the appearance of moss. Moss agates sometimes contain sections of well-organized fortification banding.

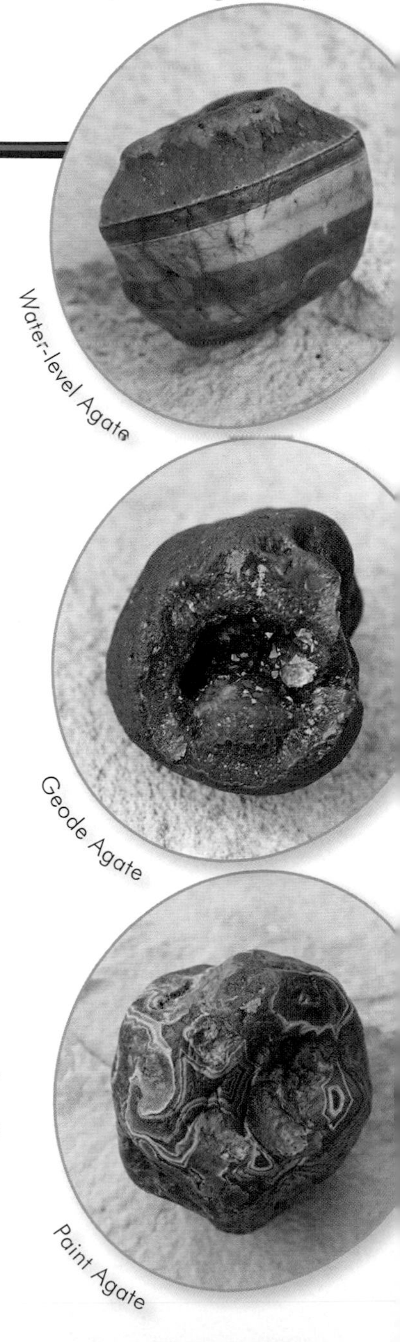
Water-level Agate

Geode Agate

Paint Agate

## Water-level Agate

- Agates with either completely or partially parallel bands. It's theorized that the material in these agates may be heavier than in fortification agates and thus the bands formed at the bottom of the vesicle in the host rock.

## Geode Agate

- Agates with a hollow center that often include crystals in their interior. These crystals may be white or gray quartz crystals, purple amethyst crystals, or dark smoky quartz crystals.

## Paint Agate

- Agates that contain a broader range of colors, including bright orange, pastels (blue, yellow, pink, rose, green), and deep/rustic brown. These agates also have some of the most intricate banding.

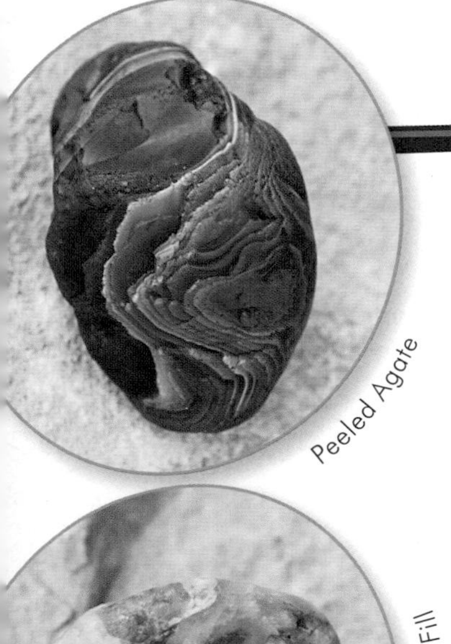

Peeled Agate

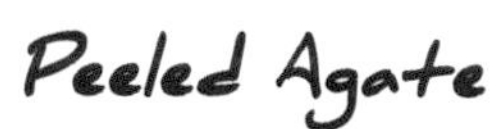

## Peeled Agate

- Peeled agates are fortification agates with layers that have been "peeled back" due to the weathering of wind, water and ice.

Amethyst and Smoky Quartz Fill

Agate in Host Stone

## Amethyst and Smoky Quartz Fill

- Agates that have a portion of the stone filled in with purple-colored amethyst crystal or dark black smoky quartz. These agates are rare and the amethyst-filled stones are especially prized.

## Agate in Host Stone

- Agates that are fully embedded in basalt, which is the core material of the Lake Superior lava flows. Basalt is either dark gray or brown and the agates will usually be red, white and pinkish stones protruding from the basalt.

Oxidized/Bleached

Quartz Ball

Floater Agate

## Oxidized/Bleached

- Agates that have been exposed to the sun for a long period of time have lighter or whitish bleached sections that may also appear somewhat iridescent. These bleached highlights usually enhance the beauty of Lake Superior agates.

## Quartz Ball

- Agates that have more than 50% quartz fill are often referred to as quartz balls. These agates usually have the darker red or gray outer husk, and maybe a few banding layers, but are then entirely made of quartz. These agates have little or no monetary value, but they are beautiful in a Lake Superior rock garden.

## Floater Agate

- A floater agate is a silica or quartz-ball agate that has a fortification agate fully surrounded by the host quartz material.

Seam Agate

## Seam Agate

- Seam agates have horizontal lines of gray or white amid chalcedony that is black or a darker gray. The most beautiful seam agates are jet black with lacy white bands. Others have more grayish hues, but the bands are always horizontal. Seam agates can also be cut and polished.
- Lake Superior water-level agates also have horizontal bands, but they have straighter and thinner/tighter lines than the banding present in seam agates.

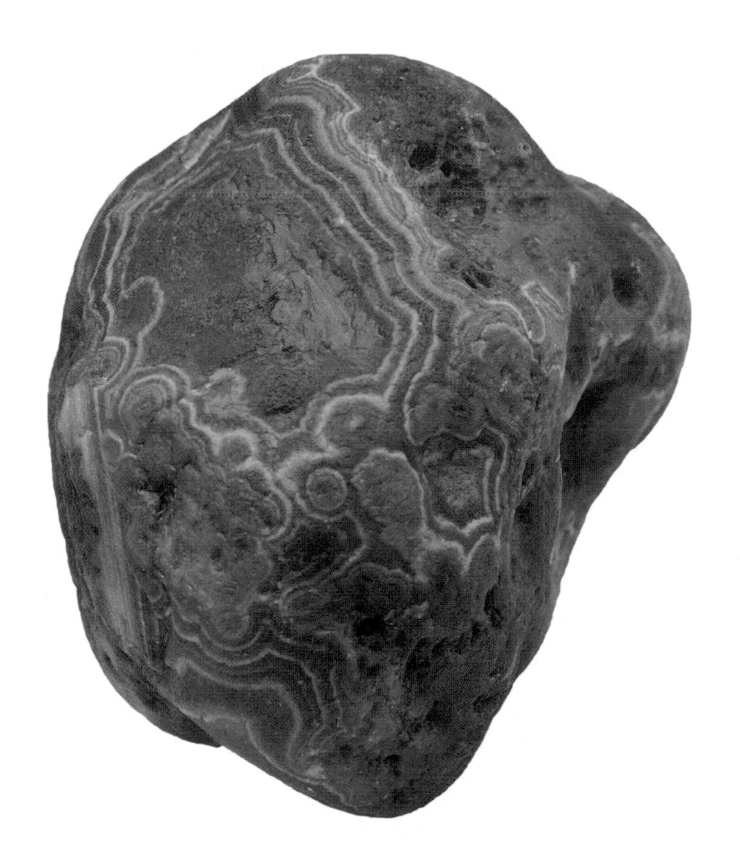

A fortification agate with eye inclusions

It is important to know which stones are not agates. And while these might not be what you are after, many are beautiful in their own right.

# Agate Imposters

## Lake Superior Gravel Identification

By now, you are getting anxious to start the hunt. In the previous section, you learned about the main characteristics of Lake Superior agates, but as the old saying goes, "A little knowledge is a dangerous thing." Too often, would-be agate collectors go agate hunting without learning about the many agate imposters that exist. It is precisely at this point that most people get stalled out and frustrated; they come home with lots of stones, but no Lake Superior agates. To avoid this frustration, it's key to be able to identify Lake Superior gravel. Remember: **When you find Lake Superior gravel, you are in the presence of agates.** Once you find Lake Superior gravel, you're halfway home, but you still need to know **what's not an agate**. Agate look-alikes, often called "foolers," are confusing, either because their coloration resembles that of a Lake Superior agate, or because they exhibit banding patterns.

In this section, we've included the stones you're likely to mistake for a Lake Superior agate. Here's one more major tip for success: Assemble a group of these "imposters" so you can compare them to the real thing. You may still decide to collect some of these finds, but you'll do so by choice, not because you mistakenly think you've found an agate.

## Chalcedony

- Because of chalcedony's waxy glow and dark reddish/purple coloration, you are sure to pick a few of these up as you learn about agates. Some chalcedony specimens have an extensive array of color variations that can be quite vibrant—these are great pieces for your Lake Superior rock garden!
- Most chalcedony pieces have flat surfaces and angular corners, but agates usually have rounded corners. Chalcedony also never has agate banding, tubes, eyes or other agate patterns. (That's actually the technical difference between them; agates are the banded variety of chalcedony.)

## Jasper/Banded Jasper/Jaspelite

- These imposters go by several names, including jasper, banded jasper, jaspelite, and even banded iron. In this text, we will refer to all of these as jasper. Jaspers come in numerous color and pattern variations but almost always have some reddish coloration and some banding. These stones can be cut and polished just like agate and can make beautiful jewelry pieces. Some banded jasper contains magnetite, a magnetic mineral. Magnetite usually occurs in black or gray bands—you can determine if your jasper has magnetite with any magnet.

- Unlike Lake Superior agate, jasper often has flat surfaces and angular corners. Also, jasper banding is not nearly as distinctive as agate banding, and you'll rarely see pits or staining on jasper specimens.

Jasper

Banded Jasper

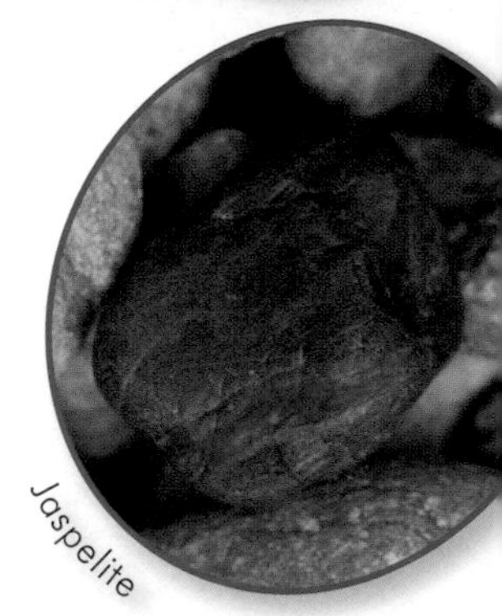
Jaspelite

Chert

## Chert

- In Lake Superior gravel, chert is generally a chalky white color and may exhibit alternating gray and white bands. Because it consists of quartz, chert also gives off a waxy glow.
- Chert does not have the striking, well-defined patterns common to Lake Superior agates, and you'll never see chert with the reds and oranges common in Lakers. Lastly, chert is always opaque, while almost all agates are translucent. (Paint agates are opaque, but they are brilliant orange.)

## Banded Flint

- Technically a variety of chert, banded flint exhibits varied color bands, making it possible to mistake it for an agate. Nevertheless, these color variations are usually black and brown with some occasional whitish bands. The banding is always straight and horizontal and not nearly as distinctive as the horizontal banding on Lake Superior water-level agates. Finally, flint has flat, angular surfaces and doesn't have agate's hard waxy/glassy surface.

Banded Flint

## Vesicular Basalt and Amygdaloidal Basalt

- Basalt is the most common of all Lake Superior gravels. It is the host stone in which Lake Superior agates form and is generally dark gray in color but can also be brown. There are two primary types of basalt. Vesicular basalt has surface pits that are empty. Amygdaloidal basalt has pits that have been filled in and contain pink, white or green crystal inclusions.
- The surface pits in basalt will often get your attention, but you'll quickly learn to differentiate these from agate pitting, because of basalt's dull gray-brown colors and its lack of other distinguishable agate features, such as banding or a hard, waxy surface.

## Granite and Gabbro

- Granite comes in a stunning variety of color combinations and exhibits something of a "salt and pepper" appearance with a variety of intermixed black, white and pink crystals. Granite with larger pink quartzite crystals is visually appealing and will catch your eye on a sunny day. Granite has a look-alike called gabbro that is strictly black and white and has a coarser surface than granite. Granite doesn't exhibit any of the standard Lake Superior agate patterns and also doesn't show pitting or staining.

Granite
Gabbro

## Amygdaloidal Rhyolite and Flow Banded Rhyolite

- Amygdaloidal rhyolite is a peach- or orange-colored stone with very small white or light blue crystal inclusions that are distinct from the host rhyolite. One of the most common "fooler stones," it is plentiful, has just enough of a reddish hue to get your attention, and also has some pitting.

- Amygdaloidal rhyolite doesn't exhibit banding or any other common agate pattern. It is not as hard as an agate, nor does it have its waxy luster. In addition, Lake Superior agates do not have small crystal inclusions like rhyolite does.

- Flow banded rhyolite is a peach or orange colored stone with a wavy banding pattern around the entire stone. Because of the orange coloration and its wavy bands you will initially be drawn to them. Its banding patterns are not nearly as distinctive as an agate's and the stones don't have the hardness or waxy luster of an agate. These stones also usually have flat and angular surfaces.

## Porphyry

- There are numerous variations of porphyry and many are beautiful. "Snowflake" porphyry stones have feldspar crystals with shapes that look like snowflakes. Because porphyry often has reddish crystal inclusions and striking patterns, it will catch your eye.
- Porphyry is easy to differentiate from Lake Superior agate because it has a grayish or white host stone and its surface is rougher than an agate's hard and glassy exterior. Also, porphyry will not include agate-like features, such as banding, tubes or eyes.

Porphyry

Porphyry

Sandstone

## Sandstone

- Sandstone comes in a variety of colors and may or may not have patterns. Many of the colors and patterns are striking and beautiful and some of the red and white banding will get your attention. Even so, banding patterns in sandstone are thicker and much less distinctive than agate banding.

Thankfully, sandstone is easy to identify because of its coarse, sandy texture.

## Conglomerate

Conglomerate

- A form of natural concrete, conglomerate contains multiple fragments of various types of stone. This varied appearance might lead one to mistake it for a ruin agate or a brecciated agate, which also consist of fragments of stone. Nevertheless, agate banding is visible in ruin agates and brecciated agates; like puzzle pieces, pieces of the banding appear to break off in one place and then resume at a different place on the stone.

- Ruin agates and brecciated agates are quite rare. Conglomerate is far more common and doesn't have a hard waxy/glass surface like agates do. Also, in ruin agates and brecciated agates, you can still see agate banding, though segments are jumbled and not continuous.

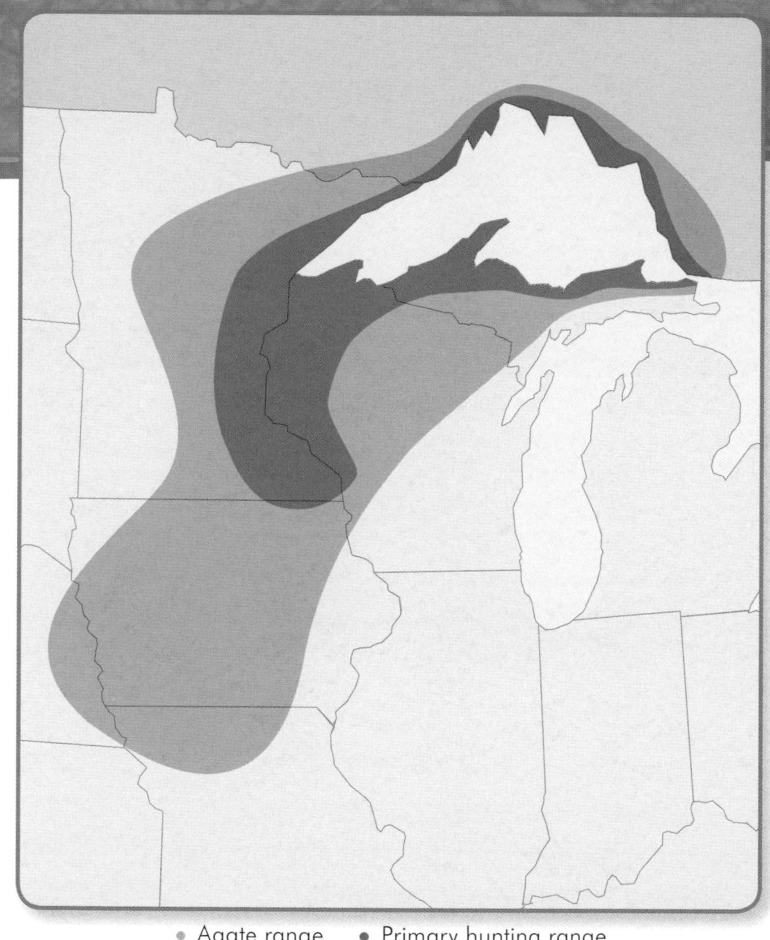

• Agate range • Primary hunting range

Hunting for agates starts with knowing the geographic range of the Lake Superior agate.

# Where to Find Agates

## Location, Location, Location

Since you know what to look for (and what to avoid), it's time to find a place to hunt for agates. It all starts with knowing the geographic range of the Lake Superior agate, which is actually quite large. The simple rule of thumb is: the closer you are to the primary hunting location, the more likely it is you'll find agates. By definition, this approach is imprecise, but this book is all about increasing your likelihood of success and helping you focus your precious hunting hours!

The map shown here reflects the primary range for finding good quality Lakers. However, Lake Superior agates can be found significantly outside of this range, even as far off as Iowa and Nebraska (where some very large "trophy agates" have been found). Agates can also be found in a narrow corridor along the Mississippi River all the way to the Gulf of Mexico.

Once you are in the right geographic area, you need to zero in on locations with lots of Lake Superior gravel. This often requires some research and legwork, but if you're serious about making agate hunting a regular hobby, it's well worthwhile. Your efforts will pay dividends—in agates! When you set out to start hunting, keep the rock hunter's general credo in mind: Only hunt where you have permission to collect agates. If you're visiting a state or national park, be sure to check with park staff first to see if rock collecting is allowed. Some parks allow collecting for free, while some charge a fee, and others allow no collecting whatsoever. **In addition, always get permission before hunting on private property, use good judgment and obey posted signs.** Our experience has taught us there is an even more important credo—**be truthful**. If you have previously been given permission to hunt, and someone stops to check in on you, keep your story straight and let them know who gave you permission. If they still say "no," it means no. It sometimes helps to tell people your name, where you're from, and that you're just out to enjoy a little agate hunting and that you are always respectful of property.

*Keep in mind that if you keep your hunting location secret, it's worth its weight in high quality Lakers!*

A trophy find

# Gravel Pits

**Accessibility:** Most difficult to gain access.

**Probability:** Excellent probability of finding agates.

**Finding the gravel pits:** Google Earth is a great tool for finding gravel pits, but not for identifying pit owners and operators.

- Go to earth.google.com and download the free version of Google Earth. Once you start the application, zero in on the state and county where you want to locate a gravel pit. Take note of obvious features that are not gravel pits, especially cities/subdivisions, wooded and grassy areas, lakes and rivers, and farm fields, which are represented by obvious square/rectangular areas.
- Then, begin to identify gravel pit "targets." Look for odd-shaped tracts of land that are grayish or sandy in color. Gravel pits could possibly be near water; deep excavations often lead to large pools of standing water.
- Zoom in to a level where you can distinguish gravel pit features. Specifically, look for cone-shaped piles as well as gravel-mining equipment, such as long conveyer belts, loaders and dump trucks.

## Example

Practicing in Google Earth is helpful, so let's walk through an example. Begin by spotting an odd-shaped tract of land just to the east of what appears to be an airport runway. It is sandy colored, so that's a good sign, and there is possibly a small pool of water nearby, which is also promising.

Taking a closer look, the sand and gravel piles stand out more and it's clear there is definitely a small pool of water present. At this point it's clear that it's not farmland. In fact, it looks like there may be some piles of sand and gravel. From this, it is becoming clear that this is a gravel pit.

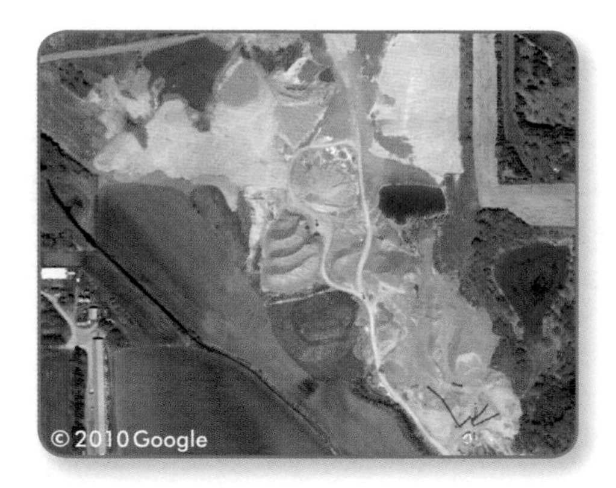


Now you can see there is some sand and gravel mining equipment on-site. See those long dark lines? Those are the conveyers used to sort materials onto different piles (various sand gradients, and multiple sizes of gravel).

Zooming in, this looks like the active gravel pit that all Lake Superior agate hunters dream about. Once you've found a gravel pit, it's time to find out who owns it (see page 116). Then, make a connection and see if there's someone that's willing to let you hunt during non-business hours. If you are serious about this hobby, you'll be willing to take a few rejections, and you'll be well advised to make it a well-kept secret once you are given approval to hunt at a pit like this one!

## Other Ways to Locate Gravel Pits

- Read through the *Minnesota and Wisconsin Atlas and Gazetteer.* These maps often contain symbols to designate locations of gravel mining operations but they don't identify the pit operators, which you'll have to determine on your own.

- Use publicly available resources to identify local gravel pits or gravel pit operators. You can also use the local yellow pages to identify aggregate and road construction companies. For your convenience, we've included some helpful resources on page 116.

- Word of mouth: Stop by a local provider of landscaping stone or of excavating services and ask who their suppliers are. See if they can provide you with a contact person for the supplier. Contact the supplier and see if there are any gravel pits that they know of that allow recreational agate hunting.

## How to Gain Access Once You've Found a Gravel Pit

- For larger commercial gravel pits, identify the pit owner/operator and call the company that operates the pit. If you know someone that works at the company, that is your best chance of getting approval to hunt in the gravel pit. Use your networking skills to make connections and assure owners that you won't hunt during hours of operation.

- For smaller/private gravel pits, stop at a nearby house to find out who owns the gravel pit and then make a connection with the owner to ask permission to hunt.

- Stop by the pit when there are operators present and talk with them about the pit policy; specifically, ask whether you can hunt when they aren't working. Or possibly ask for the owner's contact information. A small "donation" might help things along, but it's usually not necessary.
- While many people seek to hunt in the sand walls of a gravel pit, many forget about new road cuts, commercial excavation sites, and steep embankments along rivers and lakeshore. These are good places for success as you will probably be the only person to get your eyes on this gravel!

## Hunting method for rock piles

- Inspect stones for size. The best hunting piles contain stones that are between quarter size and golf ball size. See the agate size chart on page 30 for more details on why you want to start here.
- Inspect stones for condition. Look for whole stones; piles with too many stones that have been broken and crushed in the gravel-mining process will yield poor quality agates.
- Look for piles with lots of bright colors and a variety of the Lake Superior gravel types discussed earlier (see page 41). Piles dominated by gray and brown basalt stones probably have a lower proportion of Lake Superior agates.

- Check for footsteps on the pile. Piles that don't show any evidence of prior hunting/climbing are best. If you're hunting on a well-climbed pile, that's OK; such piles almost inevitably yield additional agates, but you likely won't find as many agates as you would on an undisturbed pile.
- Once you've figured out where you want to hunt, begin by walking around the pile and slowly inspecting the ground near the pile. Larger whole stones often bounce off the pile and come to rest there. Walk around the pile a second time looking at stones near the base up to about 5 feet up the pile. Since there are so many stones to inspect, make sure to take your time!
- Don't climb onto the pile until you have gone around it two full times. As you start to climb on the pile, a lot of stones will come sliding down and cover up the larger whole stones at or near the bottom of the pile. On the third time around you can climb onto the pile, but transfer your weight gently as you want to avoid kicking rocks down the pile. If you're not careful, an entire side/face of the pile can slide down, covering up nice, larger whole stones. As you go around the pile, ascend in a slight upward spiral and watch any stones that roll past you.
- On the fourth time around, you want to create some intentional rock slides by using your foot as a rake to drag stones down the pile. Otherwise you'll proceed just as you did last time around. Always go all the way to the top, even though the stones get smaller as you go up. There are often beautiful stones right at the top.

- Now walk the base of the pile again, inspecting for stones that rolled down and/or off the pile. These are often the larger stones and they may have rolled face up!
- Now go all the way around the pile once again, using your foot as a rake to drag rocks from 3 feet up the pile down to the base. By doing this you will essentially "reset" the rock pile and get a clean face to inspect. This will allow you to climb the pile and go around one or two more times.
- You can spend 2–3 hours on a large gravel pile and walk away with 10+ nice Lake Superior agates.

A rock pile at a gravel pit

# Sand Wall Formations

**Accessibility:** Moderate to difficult in terms of gaining access.

**Probability:** Good to excellent probability of finding agates.

**Finding the sand wall formations:** There are multiple locations where sand wall formations can be found; these are described below, along with location instructions and additional information.

- Gravel pits: Sand walls are common at gravel pits. Please refer to the gravel pit section (page 56) for information about locating gravel pits.
- New road cuts and embankments: When major road construction is underway in areas with Lake Superior sands and gravels, you

can find sand walls. Areas that are especially hilly offer the best opportunity for steeper road cuts and embankments. These areas are moderately easy to access, but driving may be difficult/impossible, as roads may be closed. You may have to do some hiking or four wheeling. You also need to avoid hunting while road crews are present. Otherwise, most of the lands are public and should be generally accessible. As long as you are finding other Lake Superior gravels (including agate imposters) then you have a great chance of finding a quality Laker.

- Riverbanks and lakeshores: It is primarily the larger rivers, such as the Mississippi and St. Croix, that have hilly terrain and corresponding steep sand wall embankments. Google Earth can be a good tool to assess river geography; look for steep sandy-colored embankments rather than gentle green slopes, which are likely covered in vegetation. Once again, you'll be more likely to find agates if you are looking primarily within the shaded section of the Lake Superior agate range map (page 54). In terms of lakeshores, larger lakes, such as Lake Superior, are the best hunting locales; lakes that are nestled among hilly terrain in the primary Lake Superior range are good bets, too. Most of the areas described here are public lands and should be easy to access; however, some shorelines are privately or commercially owned, so be sure to gain permission for hunting in these areas.

- There are a few other areas where sand wall formations can be found, such as gravel pits that have ceased mining operations, or even local dumps. These areas are moderately accessible and you should always seek permission from owners and operators before hunting on these lands. Don't turn your nose up at the dumps because sometimes there is treasure in that trash!

## Hunting method for sand walls

- **CAUTION—loose sand and gravel can make sand walls steep and slippery.** Make sure that you are wearing heavy-duty hiking boots. Using a walking support to maintain your balance and footing isn't a bad idea, either.
- Like in gravel pile hunting, it is best to work from the bottom up, because a lot of sand, dirt and gravel will slide down the pile once you start climbing and that will cover the nice clean stones near the base.
- Pick a section of a sand wall that you want to hunt on and begin walking the base. Stop at areas where there are accumulations of rocks and take extra time to look through them. You will find fan-shaped washouts with beautifully washed stones. Look at stones embedded in the sand for even the faintest Lake Superior agate features.
- After you've walked the entire base, carefully climb up the hill about 5–8 feet. When the ground is soft, you'll get a bit muddy but you'll have good footing. Dry and hardened sand walls are very slippery so **TAKE YOUR TIME** as you walk across.

- Walk across the sand wall horizontally at the same height—this reduces excessive up-and-down climbing and rock/sand movement, and also provides maximum coverage. Once again, make sure to look at all visible stone surfaces, including those barely "peeking through" the sand/dirt.

- When you get to a washout gully, go ahead and climb all the way up, taking time to inspect the larger volume of cleaned stones that get deposited in these gullies. After you get to the top of the hill, descend down to the level where you were climbing across the sand wall and resume moving across it horizontally.

- Once you reach the end of the section where you are hunting, climb up another 5–8 feet and go back across. Usually you'll be near the top at this point and most sand walls have a top ledge with a lot of embedded rocks (many of them of larger sizes). Take time to look at those as well. The top ledge may not be in climbing reach so you might need that walking stick to dislodge anything that appears to be an agate, but be ready to step quickly to the side so that you don't get hit by any rocks that come loose.

# Farm Field Picking

**Accessibility:** Moderate difficulty in gaining access—mostly in finding property owners.

**Probability:** Excellent probability of finding agates but primarily in the spring after plowing is done and there have been a couple of good rains. Once the crops get over knee high, it's impossible to hunt.

**Finding the fields:** The only way to find good farm fields for picking is to get in the car and do some driving. Look for fields with lots of exposed rock on the surface. Then get out of the car to see how many stones are in the field and how large they are. One great tip is to look in the vicinity of active sand and gravel pits because they are always within a larger area of Lake Superior gravel distribution.

**Gaining access:** Stop at the house near the field you want to hunt in and ask if they are the owner and/or know the owner. Be ready to tell them how you are respectful of the land, will not litter or step on crops, and that you only want to collect a few agates, if you are lucky. If you sense resistance, you can offer them a small sum of money for the privilege of hunting on their property, but usually this isn't necessary.

## Hunting methods for farm fields

Farm fields are fantastic for hunting agates. Because of the nicely defined rows, you know exactly where you've already hunted and where you haven't. Generally the distribution of stones is uneven; some sections will be just littered with gravel and others will have none. Therefore, work the field in sections that contain the most rock. After all, there are only so many hunting hours in a day! Walk down one row and visually inspect the row you are in plus one row each to the right and left. Once you get to the end of the rocky section where you are hunting, shift three rows to the right or left and go back the other direction. While walking in this direction, look down the row you are in and one row to your left and right.

# Rivers, Streams and Lakes (Wading)

**Accessibility:** Easy to gain access; most waterways allow for public access.

**Probability:** Variable, depends on the water depth and concentration of Lake Superior gravel collecting on sandbars, rocky bottoms, outcrops or shorelines.

**Finding the best river hunting spots:**

- Check the Department of Natural Resources (DNR) website for river or lake level maps; look for low water levels. Also check Google Earth for possible access points (open lots, public parks or boat landings).

- Smaller streams with slower-moving water, especially those that are within 50–100 miles of Lake Superior, are outstanding hunting locations. For example, streams near the mouth of Beaver Bay on the North Shore of Lake Superior are loaded with beautiful agates.

**Gaining access:** Comply with "No Trespassing" postings, but otherwise, rivers and lakes are open hunting grounds!

## Hunting methods for rivers, streams and lakes

Ideally, choose waters that are no more than knee-deep—any deeper and it's difficult to view the distinctive qualities of agates. Use a walking stick (that you've measured) to determine water levels and to help with balance. Walk upstream or against the current, allowing sediment to flow behind your field of vision. In areas with strong current and lots of stones, you can stop to "rake" layers of stones away to expose additional rocks. Walk with the sun at your back to provide natural lighting as much as possible.

- Use a sand or beach scoop or snorkel gear where it's safe to do so.
- In all Lake Superior agate hunting venues, you want to pick up any stone that has an agate-like characteristic. This is especially true for river hunting because of the visual distortions caused by moving currents. When river hunting, you'll probably end up picking up 3–5 times the number of stones that you would when hunting in other locations, but the effort is well worth it, as most agates will be nice, smooth, whole stones.

## Lakeshore

**Accessibility:** Easy to gain access; most waterways are public access, but stay off of private property.

**Probability:** Very high in lakes that contain a significant volume of Lake Superior gravel on or near the shore.

**Finding the lakes:** Like finding farm fields, finding lakeshore usually involves some driving around. Once again, the closer you are to either active sand and gravel pits or Lake Superior, the better your chances. Word of mouth is also a great way to find out whether a friend's or relative's cabin is on a lake with lots of Lake Superior gravel. The Whitefish Chain of Lakes near Brainerd has a number of public shorelines with a lot of hunting-worthy gravel.

**Gaining access:** Because lakes are public lands, you have open and ready access; you just need to steer clear of private property along the shoreline.

## Hunting methods for lakeshore

When hunting on larger lakeshores, such as Lake Superior, weather conditions are an especially critical component to hunting success—and safety. First, because so many people hunt for agates on shorelines, it's best to hunt after a large storm has gone through or in the spring after ice and winter storms have replenished the stones on and near the shore. Second, if storms are brewing, **TAKE HEED**, because they can develop very quickly and can be violent. Stay out of the surf in these conditions.

If you want the best chance of finding larger stones, your best bet is to hunt near the shore, where there are plentiful stones and moderate wave action. In the case of Lake Superior, search for shoreline areas that are not protected by harbors and man-made breakwaters. These structures block the waves that bring larger-sized new rocks onto the shore.

When hunting in the water (especially on sunny days), you should wear polarizing sunglasses to eliminate glare. This is the only time that we'll instruct you to wear sunglasses when hunting for agates. In addition to the polarizing sunglasses, serious lakeshore hunters have purchased or constructed devices known as "viewing scopes." You'll need to seek these out or construct one ahead of time, as they aren't generally available. A viewing scope is a bucket or other cylinder with a Plexiglas bottom that allows clear viewing of objects on the bottom of the lake. You might also want to have a collecting scoop, so you can pick things up without having to bend over.

Once you're ready to hunt, walk into the water and stand in place. Tall rubber boots or hip waders are advisable, especially in early spring when the water is icy cold. Let the rolling waves do the work of moving the rocks; inspect what is in your field of vision. You can stay stationary in the same spot for several sets of waves.

Walk ahead 6–10 feet to create a new field of vision and repeat the process.

- For hunting in the water, you can also use snorkeling gear and a face mask. Some die-hard collectors even scuba dive. This will further improve your chances of seeing stones that others have not.
- For onshore hunting, smaller-sized agates are plentiful and hunting for these gems is a peaceful, contemplative and rewarding activity (especially for kids). Just pick a spot on the shore where there are large quantities of stones, then plop down and start sifting through them. You are almost guaranteed to find nice agates.
- Onshore hunting for larger stones will take more work, both in terms of walking and uncovering stones a layer or so down. First,

find a section of the shore that contains larger stones and do the easy work of walking along that section to see what is visible on the surface or partially visible beneath the top layer. Next, you can walk along the same section and use your foot (boots required!) as a dragging mechanism to uncover the next layer of stones. This is laborious but can be rewarding.

- If you have access to a canoe or kayak, you will be able to access remote parts of lake shorelines and find shallow waters further out from shore that have lots of premium Lake Superior gravel. This is a huge advantage over landlocked hunters, and with the exercise you get it is a win-win!

# Gravel Roads and New Road "Cuts"

**Accessibility:** Easy to gain access, most roads are public access.

**Probability:** Like gravel pits and farm fields, your best luck on gravel roads will be after a good rain as the stones get dusty and it's hard to distinguish agate features. Otherwise there is an excellent probability of finding good smaller-sized agates.

**Finding the roads:** The upper Midwest is crisscrossed with gravel roads that are lined with Lake Superior gravels. Additionally, when there is major new road construction or an expansion project, there will be large road cuts with a tremendous amount of excavation.

**Gaining access:** Since all of the roads are public, they are all open for hunting. Just make sure that you make yourself visible, and when there are road crews present make sure to keep a safe distance from heavy equipment that is being operated.

## Hunting methods for gravel roads and new road "cuts"

- For gravel roads, the best/largest stones tend to accumulate at the roadsides, and it's also safest to walk along the shoulders. That's where you should focus most of your hunting effort.
- For hunting exposed embankments and new road cuts, refer to the hunting methods for sand wall hunting (see page 64).

Any time you are in the car, leave a little time to take "the scenic route" and make note of gravel roads and road cuts that you can revisit when you have time to hunt.

## Landscaping Companies

**Accessibility:** Moderately difficult to gain access; usually you need to know the people who own the business. Just like when getting approval to hunt at gravel pits, try to make contact with the owner. Also, they might be willing to refer you to a contact they have at a gravel mining company for gravel pit access direct from the source. It never hurts to ask.

**Probability:** Landscapers get a constant supply of new stones and they are washed and sorted by size. There's an excellent chance of finding great agates at companies that do a brisk business, especially those that do large commercial projects.

**Finding the landscaping companies:** Your local yellow pages will provide you with information on local landscaping companies. A quick drive-by or phone call will tell you whether they sell Lake Superior gravel.

**Gaining access:** Try to find someone you know that is connected to the owners and ask for a referral. It also never hurts to do a little business with the company and/or to offer them a small fee for hunting through their gravel piles.

## Hunting methods for landscaping companies

Refer to the rock pile hunting methods (see page 60). You won't need to use the full method since the piles will be smaller and are less likely to have sliding rocks. Additionally, be sure to ask the owners before climbing on rock piles. Some companies will not allow this as they have concerns about liability.

To gain access, try to find someone you know that is connected to the owners and ask for a referral.

# Landscaping and Roofing Gravel

**Accessibility:** Easy to gain access; most businesses don't mind casual pickers, especially on weekends or if you are doing some business with them (e.g., staying at a hotel, eating at a restaurant, etc.).

**Probability:** High probability, especially at new businesses or those that are upgrading their landscaping with new gravel.

**Finding the landscaping rock:** As you go about your daily business, keep an eye out for businesses with new landscaping gravel.

**Gaining access:** In most cases it's not necessary to ask permission, but use common sense. For example, if the business is a bank or other type of company with significant security considerations, take the time to check in with the owner and request their permission.

## Hunting methods for landscaping and roofing gravel

This is probably the simplest hunting method. Simply section off the area you want to hunt in and walk in a grid-like pattern to ensure you are covering the whole area. Like hunting on gravel piles, you will be looking at a lot of stones, so take your time.

A specimen with intricate banding

# Seasonal and Weather Conditions for Hunting

By now, you've probably noticed that the tips we mention are meant to improve your odds of success. Some people like to refer to agate hunting as a random reward system, but if you follow all of the tips and instructions we're providing, you will be rewarded! Given the number of different places to look for agates (and the seasons/conditions one can hunt in), it's easiest to convey your "odds" for agate-hunting success in a chart.

From this chart, you can see why agates are called "lovers of light." The sun is almost always your friend when you are hunting for agates; just remember: don't bring sunglasses, and wear a brimmed hat to give some eyeshade. There are a few things that aren't intuitive from this chart, and they are covered here.

- Gravel pits usually don't start processing new gravel in earnest until late spring or early summer. Until they get rolling, you will be looking at old, picked-over materials.

| LOCATION | SPRING | SUMMER | FALL | RAIN *(before* |
|---|---|---|---|---|
| Gravel pile | • | •••• | ••• | ••• |
| Sand wall | •••• | •••• | •• | •••• |
| Farm field | •••• | •• | • | •••• |
| Lake shore | •••• | ••• | •••• | •••• |
| River | • | •••• | ••• | • |
| Gravel road | ••• | •••• | ••• | •••• |
| Landscaping rock | •• | •••• | ••• | ••• |

- After crops are a certain height, farm fields become impossible to hunt in. There is no hard-and-fast date when this happens, as it depends on when a given farmer does their planting. Picking into the summer may be possible, but it's unusual. In some fields in the fall, it's possible to do additional hunting after the crops are harvested.

- At the lakeshore, rain will give the rocks on shore a nice new shine, but picking in the water when it's raining is marginal due to rain hitting the water and the lack of sunlight. Plus, there is the threat of dangerous surf on larger lakes.

- In the fall there are often strong winds and storms that cause heavy and sustained wave action on the shores of Lake Superior. This brings up some new material for hunting. You need to be opportunistic and hit the shoreline just after the storms have blown through. Since these storms happen in fall, you can sometimes combine a fall colors trip with an agate hunting adventure.

- Some people just love lousy conditions and being "at one with nature." To that end, we'll never tell you when not to hunt (other than truly severe weather). We will simply guide you based on our experience over many years of productive Lake Superior agate hunting.

| **RAIN** *(during)* | **SUNNY** | **CLOUDY** |
| --- | --- | --- |
| ••• | •••• | •• |
| ••• | •••• | • |
| •• | •••• | • |
| • | •••• | •• |
| • | •••• | •• |
| •• | •••• | • |
| •• | •••• | •• |

# Clothing

Believe it or not, what you wear agate hunting is an important element of your success. If you are going to make one investment for agate hunting, a good pair of hiking boots is a great place to start! Other than that, you probably have everything else in your closet.

**Summer Gear**

- No sunglasses allowed; they distort color variations and blur important visual distinctions.
- A cap or brimmed hat to help block the sun.
- Heavy-duty hiking boots, steel-toed boots are optional, but helpful.
- Cargo shorts, for extra pockets.
- T-shirt, wear lighter colors on warmer days.

### Spring and Fall Gear

- Cap or brimmed hat.
- Heavy-duty hiking boots, steel toes are optional, but helpful.
- Jeans, lined jeans or long johns may be needed in cold.
- Warm vest or jacket; again, one with lots of pockets is great.
- Gloves.
- Blaze-orange colors in fall; there might be other kinds of hunters out and you want them to see you!

### In the Rain *(next best thing to hunting on a sunny day)*

- Depends on the season, also bring along a rain suit or poncho.

### River Walking /At the Beach

- Swimwear, what else, right?!
- The summer gear described, plus the items listed below.
- Closed shoes with a solid bottom; sandals allow too much sand and debris to collect and don't protect against sharp objects.
- Polarizing sunglasses for hunting in the water/surf.
- Optional items include snorkel gear (for relatively calm water conditions) and wading boots.

# Equipment and Supplies

Like any other kind of hunting, you'll want to bring along the proper equipment. But we have great news: the stuff you need for Lake Superior agate hunting is really cheap and you probably already have most of these things. People who actively pursue this hobby can easily come out ahead between what they spend on additional gear vs. the value of the gemstones they find. Of course, you'll have to pry most people loose from their favorite agates!

**Standard equipment for gravel pits, farm fields, and the lake**

- Spray bottle for water. This is the agate hunter's secret weapon! If you are lucky, you can find a water bottle pouch to hold your water bottle so you can sling it over your shoulder. It's well worth the effort to find and purchase such a product! The best place to find these is outdoor gear stores or online vendors.

- Fannie pack, front pouch, or backpack. If you will also be picking up landscaping stones (and you eventually will), go with the backpack.
- Drinking water. Don't skimp on this; on a good day you'll go through a good liter of water.
- Snacks with protein. Remember, you will often hike out a substantial distance from your vehicle; nuts or protein bars are excellent choices.
- Walking stick. Useful if you will be doing a lot of climbing.
- Sunscreen. Because of how much you'll be looking down, be sure to put plenty on the back of your neck.
- Bug spray. Optional, depending on how close to standing water or woods you will be.

**Additional water hunting gear**

- Backpack. Useful for carrying supplies that you don't want to keep dry.
- Walking stick. Useful for stability and to gauge river depth; be sure to check the depth before you step, especially in fast currents.
- Sand or beach scoop. To scoop up rocks, filter out the mud and sand, and check what's left for agates.
- Water hunting "view scope." A bucket or other tube/cylinder with a clear Plexiglas bottom for viewing stones on the lake bottom.
- Life jacket. You do not want to be caught off-guard.

Once you get your agates home, it's time to clean and take care of your finds.

# Taking Care of Your Agates

## Taking Care of Your Agates

While this book is primarily geared to teaching you how to successfully hunt for premium Lake Superior agates, we also want to teach you the basics of how to take care of your stones once you get them home. That's half the fun of collecting and it's something you can concentrate on during "the off season."

The lapidary equipment that we recommend in this section is for the amateur hobbyist. Most people start with the cheapest piece of equipment (a $30 tumbler) and that is all they ever need. Like in any other hobby, you can always purchase more equipment. As they get more serious about the hobby (and into jewelry making, for instance), collectors often gradually acquire other pieces of equipment over a long period of time. If you eventually purchase all of the equipment we recommend, it will cost you over $1,000. Even if you someday hit the $1,000 mark, this will pale in comparison to the cost of other popular hobbies, such as hunting, fishing, golfing, etc. Further, you'll be able to make considerable money selling some of your finished agates.

# Cleaning and Sorting

- "Rough clean" your agates as you hunt and collect, this way you don't end up with a lot of sand in your drainpipes at home! Most of the time this is accomplished with your spray bottle.

- At home, use a toothbrush or scrub brush with warm soapy water to clean the last dirt and sand off the stones.

- Sort your stones into four piles:

  » **Grade "A" whole Lake Superior agates.** Stones of good size and striking pattern and colors. As time goes by, your personal definition of what a Grade "A" stone is will change, but you can always turn to eBay and other Internet sites to get more objective valuations of what makes a Grade "A" stone.

  » **Medium-grade Lakers** are whole or broken stones; whole agate nodules that don't show any pattern on the outer surface can be considered "medium-grade," too.

  » **Low-quality Lakers** can be used for tumbler filler stones, rock gardens, and for your pre-agate-hunt home-work sessions.

  » **Non-agates** that you collected for lapidary and decorative purposes.

# Rating Your Finds

## Use this guide to assign value to your finds.

## Rating/Level and Features

High-quality Agates

### High Quality

- Strong color separation between bands, or other features, such as tubes and eyes.
- Bright colors, such as red, white, orange or blue.
- Unusual colors, such as purple, green or yellow.
- Rare agate types, such as eye agates, paint stones or those with amethyst.
- Minimal to no quartz fill; some "floating band" agates are exceptional, but the more quartz there is, the lower the value.
- No visible fractures, or minimal fractures.
- Tumbling, cutting and face polishing are not required to display the best features.

Medium-quality Agates

### Medium Quality

- Good separation between bands and other features.
- Bright colors; grays and "root beers" are acceptable if the patterns are distinctive.
- Less than 30% quartz fill.
- Fractures may be visible but not dominant.
- Tumbling, cutting and face polishing may be required to display the best features or fix visible flaws.

Low-quality Agates

### Low Quality

- Low color separation between bands and other features.
- Dark and muddy colors (brown, dark gray, very deep red).
- Greater than 30% quartz fill.
- Visible breaks and fractures.
- Tumbling, cutting and face polishing won't resolve the visible flaws.

# Treating and Performing Lapidary Work on Agates

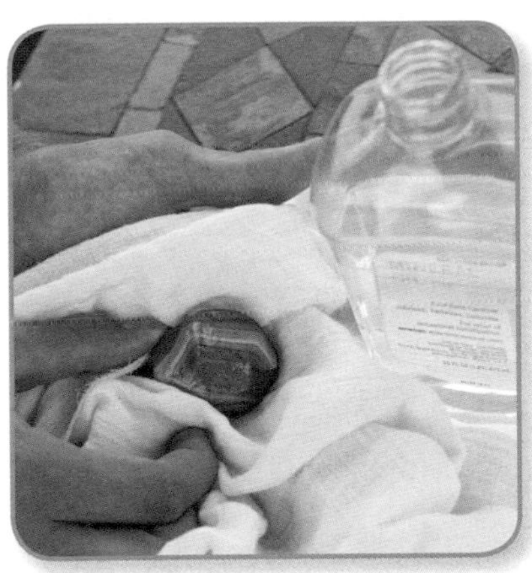
Lightly Oiled Agate

### Applying Mineral Oil to Stones for Display

- Apply a **very** light coating of mineral oil to the Grade "A" and medium-grade stones after they are dry. The mineral oil will gradually seep into the stones and "heal" any minor fractures. It will also give the stones additional luster and accentuate color separation within bands and other features.

Tumbled Agates

### Tumble Polishing

- Medium-grade stones or stones that you have cut are often the best candidates for tumble polishing. You may choose to tumble your stones to a high polish or you can perform what we like to refer to as "rough tumbling." Rough tumbling merely means that you stop after the first or second stage of tumbling and apply a light coat of mineral oil. We often suggest this method because agates often have visible fractures and mineral oil "heals" those fractures. How much to tumble your stones is really a personal judgment call.

- Don't spend money on a hobby shop tumbler. They are noisy and do not produce a good finished product. Good quality lapidary tumblers are inexpensive and can even be purchased secondhand from eBay.

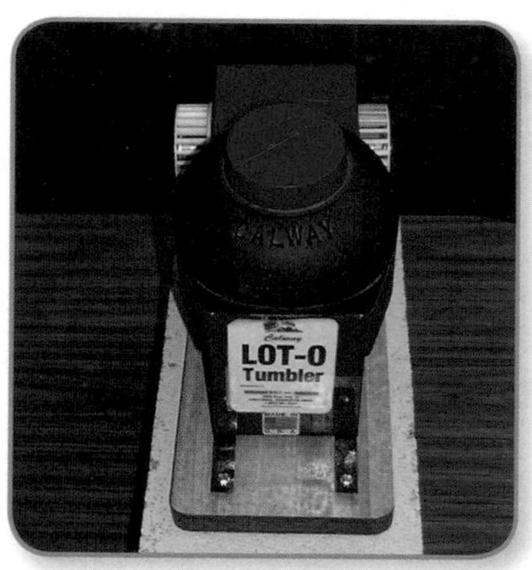

Vibratory tumbler

- There are two types of tumblers, rotary tumblers and vibratory tumblers. Rotary tumblers are much cheaper ($92 for a 3 lb. tumbler, $215 for a 12 lb. capacity tumbler), but they take up to six weeks to produce a finished product! Vibratory tumblers cost more ($210 for a 3 lb. capacity tumbler), but they get you a finished product in a week!

- We won't replicate the manufacturer's instructions on how to load the tumbler with stones, water and tumbling abrasives/grit, which are all self-explanatory. However, we will offer a few well-learned and undocumented tips below.

Rotary tumbler

- The simplest way to achieve a nice "rough tumble" finish on Lake Superior agates is to tumble your stones for 2–3 times longer than the manufacturer suggests for the first stage; then, simply remove and clean the stones, and apply a light coating of mineral oil. This avoids the hassle of tumbling (and cleaning) for multiple stages, and saves you tumbling grit.

- Build a collection of small (penny size and smaller) Lake Superior agates that you can use for buffer and filler stone. These are much more durable than ceramic and plastic fillers, and they are free!

- As you are filling your tumbler, make sure you have a good mix of stone sizes and intersperse the sizes at each level. Start by putting down a layer of small-sized stones and then mix in medium to large stones. Then, complete each layer with small agates.

Face polished agate

Lapidary sander

**Face Polishing**
*(also known as Dome Polishing)*

- When you face polish an agate, you only polish one side or part of the agate, usually the one with the most striking visible features. This may be done to preserve the beautiful natural outer "husk" of the agate and its features, such as pits and staining. It may also be done with stones that were cut because the agate had a bad break, or for a whole agate nodule with no banding features visible on the surface.
- Face polishing equipment starts at around $659 for a flat-laying lapidary sander, such as the machine made by Ameritool. This machine is perfect for amateur hobbyists and comes with a series of diamond-coated sanding wheels. You can turn out beautifully face polished display specimens in about 30 minutes. Larger commercial face polishing machines can cost $2,000 to $3,000; these larger

machines are for large-scale lapidary production of polished stones and jewelry pieces.

Lapidary saw

Polished and cut agates

### Cutting/Sawing

- A lapidary saw is an invaluable piece of equipment if you will be collecting a significant volume of agates. A six-inch diameter saw will run you about $409. Make sure to purchase a high-quality diamond-coated blade separately, which costs about $61. The diamond-coated blade will cut thousands of agates and it will make cleaner cuts, and take about one-quarter the time of the inexpensive blades that come with the saw.
- Any time you are operating cutting equipment, you should be cautious; the good news is that lapidary saw blades rarely cause even a light break in the skin, let alone a serious injury.
- You will use your lapidary saw for a number of purposes,

including cutting off broken or badly flawed portions of agates, making jewelry pendant slices, cutting open whole agate nodules with no visible pattern showing, or to get an agate ready for face polishing (see page 96).

### Sawing Tips

Here are some tips that will make your sawing activity safer, cleaner and more enjoyable.

- Make sure the area you are working in is well lit, and set up a small desk lamp to give extra lighting. With a basic trim/slab saw, you are often working with small stones and need to make reasonably precise cuts.
- As you are cutting, cooling oil will fly off the blade; this can make a big mess. Put up a piece of plastic sheeting on the wall behind where you are cutting and make a hard cardboard shield to insert behind the cutting surface. Also lay a piece of plastic sheeting on the floor. Finally, wear a cheap plastic rain poncho or old work clothes so you don't ruin good clothing.
- Saws are also noisy, so set your saw up where you are well away from other household activities. Wear noise dampening headphones or earplugs while cutting.
- Wear goggles to protect your eyes from tiny rock fragments.

Drilled agates

**Drilling**

- If you want to use your agates to make pieces of jewelry, such as pendants or beads, you'll likely need to drill into them. Many people think they can drill holes in agate slices with hardened graphite drill bits, but agate is one of the hardest materials on earth. On the measure of mineral hardness (the Mohs hardness scale), they have an overall hardness rating of seven. They are one of the hardest materials you'll find on a regular basis.

- Therefore, to drill into them, you need to have something that has been built specifically for drilling agates and other hard jewelry-quality stones. For about $245, you can acquire a small drill press that shoots a stream of cooling mixture onto the stones as they are being drilled. You also need to purchase a supply of diamond-tipped drill bits (hollow core bits are strongly recommended).

- The most critical aspect of drilling is having a soft touch, otherwise you will quickly burn off the diamond coating on the tip of the drill bit. A good soft touch will yield 10–25 drilled slices for a $2 drill bit; a rough/aggressive touch will burn out the bit in 2–4 stones!

# Displaying Your Treasures

This may be the most personal aspect of collecting Lake Superior agates. There are so many wonderful ways to display your agates. Some people like to create elaborate display boxes while others keep things very simple. Having said this, you'll want to make at least a modest effort and investment to arrange your treasures. This way, they don't become more clutter in your busy living space.

The first step is to make sure that there is abundant light where your Lakers will be displayed.

Many people will simply place a number of their best agates (fully polished or in a natural state) in a tasteful glass, ceramic or wood bowl. This is a great low-cost way to get started sharing your agates with friends and family (or with coworkers if you bring them to work). Agates make a great icebreaker as people love picking up and inspecting beautiful and natural gems. There are also a soothing influence to have around the house.

Exposed shelves and window ledges are an excellent place to set some of your premium Lake Superior agates, especially where there are bright lights or natural sunlight to get the fully beauty of the stones to "pop." You can rotate the agates you have on display regularly so they always have a fresh look.

Glass display cases can be acquired for moderate prices at gift shops and online, and there is a wide range of sizes available, depending on your space constraints. If you can construct your own display cases, that's even better. Either way, a nice glass case not only provides a beautiful display, but the stones don't get dusty and rarely need to be cleaned. The best glass cases also have embedded lighting systems. Once again, consider rotating what you have in your display case to keep things fresh.

Plastic fishing bait boxes, like those made by Plano, are a cheap and excellent way to sort, store and carry a portion of your best agates. It's great to be able to share your hobby with friends and family because most people have no idea how beautiful Lake Superior agates are. Most people start out as moderately curious, and as you continue to bring new finds over for gatherings of family/friends they become more and more interested in your hobby. Before you know it, they might be asking you to take them hunting!

In addition to tackle boxes, a medium-size plastic Tupperware box is great to display your medium-grade stones and to hold specimens you plan to give away. You can bring this along to gatherings of friends and family and let everyone pick out one or two of their favorite agates to keep. Everyone has their personal favorites and it's fun to watch everyone paw through the agates and pick out a prize.

Over the years you'll collect so many agates and other decorative Lake Superior gravel stones that you'll be well advised to create a small rock garden where they can see the light of day and enhance your yard. The best arrangement is to create a "bed of stones" and then build a layer of the Lake Superior gravel and agates on top. Then nestle and arrange flower pots within the rock garden. It is a low maintenance setup and quite attractive.

The last display setup that we'll share is for those of you that become true agate hounds. There are many people we've met that have created an "Agate Cave," the agate-collector's equivalent of a sports memorabilia "cave." These "caves" usually include floor standing display cases, shelving with additional specimens arranged on them, posters and paintings of agates (yes there are paintings of Lake Superior agates), agate art such as "stained glass/agate" windows or pictures made out of agate slices, an agate-related library, and anything else that pertains to the hobby of collecting agates.

# Buying and Selling Agates Online

Some agate hunters would never buy or sell a Lake Superior agate, and for most serious hunters, finding good quality gems is a deeply personal experience. Show them a photo of one of their prized finds and they'll probably tell you a story about finding it, including when and exactly where it was found!

Then again, an active Lake Superior agate market only enhances the hobby and opens it up for more people to enjoy. First of all, there are people living all over the world that may never have a chance to hunt for agates in the upper Midwest; for them, the online marketplace is one of the only ways for them to obtain Lake Superior agates. There are benefits for agate collectors, too; it's a great way to cover your costs for lapidary equipment and the gas money needed to drive out to hunting locations, and it can be a way to make some extra income.

Whether you're interested in buying and selling agates, it never hurts to be informed, so we're going to provide you with some up-to-date (and valuable) information about buying and selling agates online. When buying, the best advice is to do business with established sellers. When/if you start to sell your agates, commit yourself to fair pricing and advertising, and be sure to accept returns from unsatisfied customers. This just helps to make the market better and stronger.

### Top Websites

All of the websites listed below make an active market in Lake Superior agates. These sites, especially eBay, will give you a good indication of the value of your own agates.

- eBay: This is a public site where anyone can directly buy or sell agates. On any given day there are usually over 500 active listings for Lake Superior agates! This site gives you the best idea of how much you can sell your own agates for, if you have the patience to sell them individually or in select groupings. Ebay is home to numerous high-end vendors that have rotating inventories. Vendors acquire their agates from collectors after developing a business relationship with them (for more about this, see page 108).

- These vendors sell only top-quality specimens and you won't be disappointed if you buy from them. Smaller-scale sellers are usually selling items they have found themselves. You will see prices on eBay anywhere from $1 to over $2,000 for world-class "trophy" agates.

The following private sites also sell agates that have been acquired from individual collectors:

- Agatenodule.com
- Superagates.com
- Amazingagates.com

### Benchmark Prices for Private Sellers on eBay

(Agates sold on private sites are usually a bit higher.)

| Size | Quality | Price Per Ounce |
|---|---|---|
| 1–4 oz. | High | $5 |
| 5–10 oz. | High | $8 |
| 11–16 oz. | High | $12–20 |
| 16–48 oz. | High | $20–50 |
| 1–4 oz. | Medium | $2 |
| 5–10 oz. | Medium | $3 |
| 11–16 oz. | Medium | $4–10 |
| 16–48 oz. | Medium | $11–20 |

**NOTE:** You can sell larger agates that are low quality and sometimes they will go for higher than benchmark prices, but don't get your hopes up if you find a 1 lb. "quartz ball" and think it may sell for $200. It's more likely you'll get $20.

### Other Buying and Selling Tips

- When buying or selling in lots/groups, there are different types of lots/groups that are commonly sold on eBay. Make sure that you inspect these carefully if buying and that you label your own lots/groups correctly if you are a seller.
- "Tumbling rough" is a label that is commonly used to signify low-quality stones. These are sold by the pound.

- Lots consisting of specific types of agates, such as eye agates or paint agates, can fetch a nice premium, even for smaller stones. These are great for "instant displays."
- Polished agates are often sold in lots/groups and are usually medium-grade stones. These are also great for "instant displays" or for giveaway items at children's events.
- High-grade small to medium agates are great to purchase if you are interested in creating Lake Superior agate jewelry. Such stones may also be used for display.
- If you're going to try to sell your agates, dress them up. Imagine if you were searching online for a car and you found a listing with a grainy photograph where you couldn't even exactly tell what the make and model of the car was, let alone the condition it was in. Most people would just skip past that ad unless there was some ridiculous low price. Even then, they'd probably be suspicious. The most successful vendors take nice, high-resolution photos and use catchy and descriptive text to grab buyers' attention and give them all the information they need to make a buying decision.

Only purchase Lake Superior agates from top-end vendors. When buying keep the following tips in mind:

- Look for vendors that only sell high-quality agates, i.e., they don't sell a mix of poor, medium and high-quality stones. In addition to selling individual agates, they sell some nice collections of agates, including groupings such as eye agates, "paint" agates and polished agates.

- Don't buy from vendors that have a customer satisfaction rating lower than 98%. Our preference is to only buy from vendors who have 99–100% customer satisfaction. If you are buying a large/high-priced Laker, this becomes even more critical.

- The more completed sales a vendor has, the more reliable their product is. Vendors who also have their own eBay stores tend to be more reliable.

- Final note on selling: If you don't want to go through the trouble of selling your own agates individually, the larger top-rated vendors will purchase agate collections, but they will pay wholesale prices (about 50% less than what we show in the agate value chart). Remember, they need to make a profit and they have to cull through a lot of agates before finding ones they want to photograph, list, sell, ship and pay seller's fees and taxes.

## Some Real-World Examples

To give you an idea how agate sales go in the real world, we're including information about a few recent agate sales conducted online. This way, you'll get a very rough idea of what to expect when selling your agates. We've provided information about each stone, including the title of the original online sale, the weight of the stone, and the realized price for each stone.

Sold for $389

### Lake Superior Agate— Top Shelf 1 lb., 1 oz. Trophy

**Notes:** A high quality stone with a distinctive and bright white fortification. Weight: 17 ounces.

**Selling price:** $389, or $23 per oz. It sold on the lower end of the average price range for a stone of this size and quality, but it's a fair price.

Sold for $108

### Colorful 13 oz. Bander— Nice Collector's Piece

**Notes:** A nice size whole stone, but pattern doesn't show much contrast. Weight: 13 ounces.

**Selling price:** $108 or $8 per oz. This sold in the middle of the price range for a stone of this size and quality.

Sold for $38

### Stunning Lake Superior Floater Agate

**Notes:** The bright and thick white band on this stone is beautiful, but there is too much uninterrupted quartz. Also note that the stone doesn't have floating bands, so the advertised title is incorrect. Weight: 11 ounces.

**Selling price:** $38 or $4 per oz. This sold at the bottom end of the average price range for a stone of this size. This is what too much quartz can do to pricing.

Sold for $77

### Perfect Candy Striped Laker with Awesome Triangle Pattern

**Notes:** The color contrast and striking pattern make this a classic agate. Weight: 6 ounces.

**Selling Price:** $77 or $13 per oz. This sold on the high end of the average price for a stone of this size, but that is well justified based on the premium quality characteristics.

Sold for $25

### Beautiful Lavender Shadow Agate

**Notes:** A nice-sized whole stone but with very little color separation in the banding. Weight: 6 ounces.

**Selling Price:** $25 or $4 per oz. This sold at the average price range for a stone of this size and quality

Sold for $67

### Collector's Grade Water-level with Unique Blue Bands

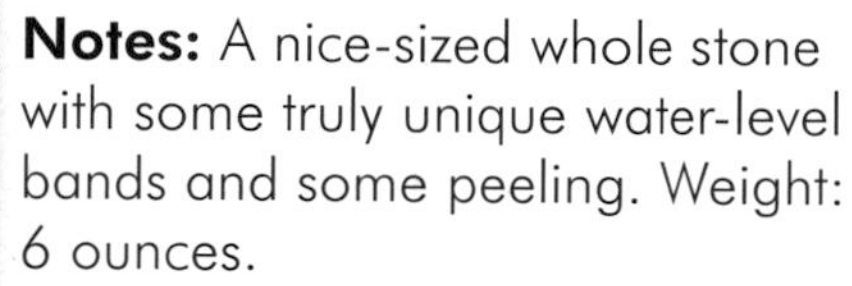

**Notes:** A nice-sized whole stone with some truly unique water-level bands and some peeling. Weight: 6 ounces.

**Selling Price:** $67 or $11 per oz. This sold above the average price range due to unique features and overall quality. The title of this stone brought the valued features to the buyer's attention.

Sold for $83

### 1 lb., 11 oz. of Premium Agates

**Notes:** A beautiful selection of smaller-sized Lakers

**Selling Price:** The lot totaled 27 ounces with just over 1 oz. per stone. It sold for $83 or $3 per oz., so this sold below the average price range for agates of this size and quality. These kinds of collections are highly variable in what they fetch; it's common to see a nice collection like this go for well over $100. Always watch for accurate labeling; lower-quality collections should be labeled as rough/tumbling stones.

# Glossary

**Lake Superior gravel** There are numerous types of stone that can be found on the shores of Lake Superior, including sandstone, basalt, granite, and, of course, Lake Superior agate. During the various periods of glaciation, large amounts of these stones were spread across the upper Midwest, sometimes in concentrated deposits. Gravel-mining companies extract these gravels and use them for many purposes, including road construction, concrete production and landscaping.

**Fooler/Imposter** Many of the stones found in Lake Superior gravel deposits exhibit one or more of the characteristics that are common to Lake Superior agates, such as banding, reddish colors, waxy glow or pitting. As such, it is common for beginners (and even veteran hunters) to be fooled by these "agate imposters." As time goes by, it becomes easier to identify these "fooler" stones and one is less likely to pick them up for inspection.

**Landscaping gravel** Landscaping gravel is mined from commercial gravel pits, sorted to various sizes, and washed. Essentially, it's the same thing as Lake Superior gravel. Landscaping gravel is used for decorative purposes around homes and commercial businesses. It is also used as roofing gravel to form a protective barrier on top of roofing tar.

**Tumble polished agates** Agates that are polished via tumbling in a machine that rotates or vibrates. This causes the stones to be worn down to a smooth and/or glossy finish. Tumbling "grit" is used to accelerate the polishing process, which begins with a tumbling cycle with coarser grits, followed by tumbling cycles with finer and finer grits. A final tumbling cycle applies polishing compounds to the ultra-smooth stones. Not all stones are good candidates for tumble polishing; because agates are prone to fracturing, it is sometimes preferable to tumble the agates smooth but skip the polishing cycle, and then apply a light coat of mineral oil rather than polishing.

**Face polish** In an agate that is "face polished," only one surface or "face" of the agate is polished, usually the surface that has the most striking pattern. This leaves the rest of the stone in its natural state. Face polishing is a process similar to tumbling, in that a series of grinding wheels with varying levels of coarseness successively wear one surface of the agate down to a smooth or polished state. Face polishing an agate takes 20–30 minutes, whereas tumble polishing takes weeks.

**Sand wall** A sand wall formation is a large deposit of glacial sand and gravel that has been either commercially excavated or weathered out by water and wind. Sand walls are usually steep hills at an angle of 45–75 degrees and can be anywhere from 10–100 feet tall. Sand wall hunting requires agility, caution and stamina, but the techniques we describe in this guidebook will help you successfully hunt in sand walls. Not everyone is up for sand wall climbing, so it offers a great opportunity to be the first person to see stones that have been buried for over 10,000 years.

**Gravel pit** Commercial sand and gravel mining operations are commonly referred to as gravel pits. These pits can range in size from an acre to several square miles. Gaining access to these gravel pits is difficult because of legal liability concerns, but the effort to gain access pays lasting rewards.

**Fortification** Whether they are Lake Superior agates or agates from a far-flung corner of the globe, most agates exhibit concentric rings/bands that are referred to as a "fortification" structure. This structure gets its name from its resemblance to old-fashioned forts that had concentric walls that served as successive lines of defense against invaders. In a fortification agate, the concentric rings/bands can be very smooth and almost circular, or they can be highly intricate. Also, the rings can be very wide or very narrow/thin, and color combinations across these ring patterns are highly diverse. An agate's beauty and value depends a great deal on the intensity and color variations of the fortification rings.

**Husk** The outer skin of an agate is often referred to as its husk. This term refers to the rough and rumpled appearance that is common to most agates in their natural state. Some people say that rough agates look like old potatoes, so if you see a stone exhibiting that type of appearance you may well be looking at an agate.

**Vesicle** Small bubbles in host stones are referred to as vesicles. These air pockets formed in lava masses as they cooled; such lava masses underlie much of the Lake Superior Basin. Later, some of these vesicles filled in. In some, agates formed, while various crystalline materials filled in others. Some simply remained empty and appear as spherical cavities in the host stones.

**Opaque** Describes material that doesn't allow light to pass through it. The vast majority of agates are translucent, and light passes through them. However, some agate varieties, such as "paint agates," are opaque. Paint agates also exhibit the greatest range of colors, and the colors themselves are more striking/vibrant, especially the orange hues.

**Translucent** Describes material that light can pass through or penetrate. Most Lake Superior agates are translucent and this is one of the primary characteristics to look for when agate hunting. This is also why sunny conditions are quite helpful for successful agate hunting.

**Inclusions** Air pockets in a host stone, such as basalt or rhyolite, often fill in with other materials. Such materials are called inclusions. The material that forms in these air pockets is often crystalline, but there are many other types of minerals that can form these inclusions.

**Chalcedony** Chalcedony is a very hard silica-based mineral and Lake Superior agates are a banded form of chalcedony. Unfortunately for agate hunters, several stones that are commonly found in Lake Superior gravel are very similar to chalcedony. All of these look-alikes exhibit chalcedony's waxy luster and can resemble agate in terms of coloration. These look-alikes are what we refer to as fooler or imposter stones. One of the most significant characteristics of these "garden variety" look-alikes is that they have flat, angular surfaces, whereas agates generally have more rounded surfaces.

**Lake Superior agate** Dozens of agate varieties are found in different geographic locales around the world. Lake Superior agates get their name because they formed in the lava masses that make up the Lake Superior Basin. Beloved around the world, Lakers are Minnesota's state gemstone.

**Rock** Simply put, rocks are combinations of minerals. Rocks can form in many ways. Some rocks form quickly and violently (as in a volcanic eruption), while others form over the course of millions of years. Worldwide, there are hundreds of different types of rocks, all with different mineral compositions. We like to say that every rock has a story to tell because of its varied creation and distribution by natural forces (such as glaciation or erosion). Lake Superior agates usually formed in basalt, a volcanic rock.

**Mineral** Minerals are chemical compounds. Rocks, on the other hand, are a mixture of minerals. While rocks usually form during large-scale geological events (such as an eruption), minerals usually form via chemical processes. As agates consist of a silica-based compound, they are classified as minerals.

**Crystal** The term crystal can either refer to compounds that form various kinds of minerals, such as quartz, or it can refer to a structure with geometric shapes. Crystal inclusions occur in many Lake Superior gravels, and crystal structures that were already present when an agate formed can leave sharp angular pitting features or "dimples" on the exterior of the agate.

## Reading References

Brzys, Karen. *Understanding and Finding Agates.* Grand Marais: Gitchee Gumee Agate and History Museum, 2004.

Carlson, Michael. *The Beauty of Banded Agates.* Fortification Press: 2002.

Lortone Incorporated. *Professional Gemstone Tumbling.* Mukilteo: Lortone, 2004.

Lynch, Bob and Lynch, Dan R. *Agates of Lake Superior: Stunning Varieties and How They Are Formed.* Cambridge: Adventure Publications, 2011.

Lynch, Bob and Lynch, Dan. *Lake Superior Rocks and Minerals.* Cambridge: Adventure Publications, 2008.

Pabian, Roger. *Agates—Treasures of the Earth.* Richmond Hill: Firefly Books, 2006.

Smith, Edward. *How to Tumble Polish Rocks into Gems.* Secrets of the Pros Press, 2007.

Stensaas, Mark. *Rock Picker's Guide to Lake Superior's North Shore.* Duluth: Kollath-Stensaas Publishing, 2003.

Wolter, Scott. *The Lake Superior Agate.* Eden Prairie: Outernet Publishing, 1999.

## Helpful Resources

As mentioned in the gravel pit hunting section, you need to do your homework on gravel pit access well ahead of your planned outings. Remember that finding a gravel pit is just the beginning of the process. Once you have found an active pit, you need to make connections with the owners and operators to get authorization to hunt. The resources shown below are the best published information we are aware of. For other states in the Midwest, we suggest looking for aggregate mining resources; these are often available through a state's Department of Transportation website.

One additional tip: rock clubs often are able to arrange field trips to active pits for members; this is also a great way to meet other agate hunting enthusiasts.

### Minnesota

- A detailed, interactive map of gravel mining locations, great for finding pits, but not for determining who owns and operates them: www.mrrapps.dot.state.mn.us/gisweb/viewer.htm?activelayer=8

- A general list of aggregate and road construction companies, most of which also operate gravel mining operations: www.asphaltisbest.com/memberlist-contractors.asp

- A Minnesota Department of Natural Resources map that shows where there are high concentrations of Lake Superior gravels. This resource is great not only for finding possible gravel pit locations but also where there are likely to be farm fields, lakes and streams that contain high concentrations of agate-rich sediments: www.dnr.state.mn.us/lands_minerals/aggregate_maps/completed/index.html

- A Minnesota Department of Transportation map showing county gravel pits, rock quarries, and commercial aggregate sources by county: www.dot.state.mn.us/materials/aggcopitmaps.html

### Wisconsin

- A general list of aggregate and road construction companies, most of which also operate gravel mining operations: www.aggregateproducers.org/documents/ProducerMembersbyCounties2011.pdf

## About the Author

**Jim Magnuson**

Rock hounding is more than a hobby for author Jim Magnuson, it's a serious and rewarding avocation that helps him connect with nature. He has been an avid hunter and student of various gems, minerals and fossils since his childhood, when he first began to hunt for stones in his native state of Illinois. In addition, Jim enjoys sharing his passion not only through showing and gifting some of his finds, but also through writing, another lifelong interest. Throughout Jim's career as an Information Technology professional, he has developed his technical writing skills while creating new processes that reduce complexity and improve efficiency. These same skills proved to be invaluable as he set out to create a modern-day guide for beginning agate hunters. Jim is also a member of the Minnesota Mineral Club and enjoys attending other rock and mineral clubs as a way to further his learning and branch out into other types of agates, gemstones and geology.

## About the Photographer

**Carol Wood**

Carol Wood took up professional photography as a means of satisfying a lifelong passion for creating and sharing things of beauty. She has a keen eye for seeing perspectives in things that on the surface appear to be mundane or quite simple. Given her training and natural instincts for perspective and complementary lighting that enhances visual clarity, Carol is able to produce high-definition photographic images that enhance but don't distract from the given subjects. These skills are essential in providing a guidebook that novice agate hunters can use as a just-in-time visual reference guide. In addition to Carol's photographic pursuits, she also enjoys outdoor activities with her friends and family, especially activities that have both a mental and physical component. As a result, she has become an avid rock hound in her own right and has gradually built a collection of beautiful agates that adorn her home in northern Illinois. Carol's personal interest in the hobby has helped her to walk in the shoes of the beginning rock hound and thereby envision and create a photographic learning experience.